Amber has two big bags of food.
Two is a good number.

"Look at this food,"
said Amber.
"I love a picnic!"

Mom has four cans of punch.
Mom sets the punch here.

"I see an ant," said Amber.

"We need one ant at a picnic."

Dad gets nine sacks from a box.
Nine is a good number.

"We will have fun," said Amber.

"My pals will be here soon."

Two, four, six, eight pals run from a van.
Amber and her pals will get into the sacks.

Eight is the best number!
Eight plus one have fun.
Who will win?

The End

Understanding the Story

Questions are to be read aloud by a teacher or parent.

1. What is the title of this story? (Eight Is the Best Number!)

2. What are Amber and her parents doing? (getting ready for a picnic)

3. How many friends come to the picnic? (eight)

4. What do you think everyone will do after the sack race? (Possible answer: eat and have fun)

Saxon Publishers, Inc.
Editorial: Barbara Place, Julie Webster, Grey Allman, Elisha Mayer
Production: Angela Johnson, Carrie Brown, Cristi Henderson

Brown Publishing Network, Inc.
Editorial: Marie Brown, Gale Clifford, Maryann Dobeck
Art/Design: Trelawney Goodell, Camille Venti, Jillian Gordon
Production: Joseph Hinckley

© Saxon Publishers, Inc., and Lorna Simmons

All rights reserved. No part of this publication may be reproduced, stored in a retrieval system, or transmitted in any form by any means, electronic, mechanical, photocopying, recording, or otherwise, without the prior written permission of the publisher. Address inquiries to Supervising Copy Editor, Saxon Publishers, Inc., 2450 John Saxon Blvd., Norman, OK 73071.

Printed in the United States of America
ISBN: 1-56577-961-4
Manufacturing Code: 01S0402
2 3 4 5 6 7 8 9 10 BSP 06 05 04 03 02

Phonetic Concepts Practiced

vc´|cv (number)

Nondecodable Sight Words Introduced

eight

ISBN 1-56577-961-4

Grade K, Decodable Reader 15
First used in Lesson 139

Cable Cars

written by Lucy Floyd
illustrated by Linda Bourke

THIS BOOK IS THE PROPERTY OF:

STATE_____	Book No. _____
PROVINCE_____	Enter information
COUNTY_____	in spaces
PARISH_____	to the left as
SCHOOL DISTRICT_____	instructed
OTHER_____	

ISSUED TO	Year Used	CONDITION	
		ISSUED	RETURNED

PUPILS to whom this textbook is issued must not write on any page or mark any part of it in any way, consumable textbooks excepted.

1. Teachers should see that the pupil's name is clearly written in ink in the spaces above in every book issued.
2. The following terms should be used in recording the condition of the book: New; Good; Fair; Poor; Bad.